YOU WILL PASS CALCULUS

You Will Pass Calculus

Poetry Affirmations for Math Students

Walter the Educator™

SKB

Silent King Books a WhichHead Imprint

Copyright © 2023 by Walter the Educator™

All rights reserved. No part of this book may be reproduced in any manner whatsoever without written permission except in the case of brief quotations embodied in critical articles and reviews.

First Printing, 2023

Disclaimer
This book is a literary work; poems are not about specific persons, locations, situations, and/or circumstances unless mentioned in a historical context. This book is for entertainment and informational purposes only. The author and publisher offer this information without warranties expressed or implied. No matter the grounds, neither the author nor the publisher will be accountable for any losses, injuries, or other damages caused by the reader's use of this book. The use of this book acknowledges an understanding and acceptance of this disclaimer.

dedicated to all the math lovers
across the world

CONTENTS

Dedication v

Why I Created This Book? 1
One - Calculus' Beauty 2
Two - The Spirit To Persist 4
Three - Fear Not 6
Four - You'll Conquer 8
Five - Passionate Pursuit 10
Six - Dear Student 12
Seven - Power To Create 14
Eight - Mastering Calculus 16
Nine - Success Is Yours 18
Ten - Path To Greatness 20
Eleven - You'll Surely Pass 22
Twelve - Setting You Free 24

Thirteen - Knowledge You Invest 26

Fourteen - Conquer Calculus 28

Fifteen - Growth And Achievement 30

Sixteen - Wondrous Flight 33

Seventeen - Curves And Slopes 35

Eighteen - Pure Delight 37

Nineteen - Witness The Miracles 39

Twenty - Self-transcend 41

Twenty-One - Reach For The Skies 43

Twenty-Two - Understanding Calculus 45

Twenty-Three - Expand Your Horizons 47

Twenty-Four - Boundless Sea 49

Twenty-Five - Brilliance Expressed 51

Twenty-Six - Possibilities And Paradox . . . 53

Twenty-Seven - Your Mind Unfold 55

Twenty-Eight - Perseverance And A Curious Mind 57

Twenty-Nine - Much To Gain 59

Thirty - Solve The Puzzles 61

Thirty-One - Embrace The Challenge 63

Thirty-Two - Believe In Yourself 65

Thirty-Three - Chance To Overcome 67

Thirty-Four - Rewards Of Mastering Calculus 69

About The Author 71

WHY I CREATED THIS BOOK?

Creating a poetry book to motivate students to pass the subject of Calculus was a unique and effective approach to engage their interest and help them overcome the challenges of the subject. Poetry has a way of conveying complex ideas in a creative and relatable manner, making it an ideal medium to simplify and present Calculus concepts in an accessible way. By using metaphors, imagery, and rhythmic language, this poetry book can inspire students, boost their confidence, and make the learning process more enjoyable. It can also highlight the practical applications of Calculus, showing students how it relates to real-world scenarios and their future careers. Ultimately, this poetry book can bring a fresh perspective to the subject, fostering a positive mindset and motivating students to excel in Calculus.

ONE

CALCULUS' BEAUTY

In the realm of numbers and equations, a challenge awaits,
Where Calculus' mysteries reside, opening infinite gates.
Oh, dear student, fear not this daunting art,
For within your grasp lies the power to impart.
    Calculus, the language of change and motion,
Unfolding truths that stir the depths of devotion.
With limits and derivatives, you'll chart the unknown,
And conquer the peaks of knowledge yet to be shown.
    Through integrals, you'll unravel the secrets of space,
As derivatives guide you to a higher plane's embrace.
From rates of change to optimization's quest,
Calculus unveils the world's hidden zest.

Embrace the challenge, let curiosity ignite,
For in the depths of numbers, lies a boundless flight.
Expand your mind, let Calculus be your muse,
And witness the wonders it shall infuse.

In every theorem and proof, wisdom shall unfold,
As you navigate the realms where numbers behold.
Persistence and diligence shall be your guiding star,
As you strive to pass this subject, no matter how far.

So, dear student, let not discouragement take its hold,
For Calculus' beauty is a tale yet untold.
With determination and focus, you shall transcend,
And conquer this subject, my dear, until the end.

TWO

THE SPIRIT TO PERSIST

In the realm of numbers, where mysteries reside,
Lies a subject known as Calculus, a thrilling ride.
A challenge awaits, a mountain to ascend,
But fear not, dear student, for I am your friend.

Calculus, the language of change and motion,
Unlocks the secrets of nature's devotion.
From derivatives to integrals, it may seem tough,
But perseverance and grit will be enough.

Embrace the beauty of limits and rates,
As you navigate curves and determine their fates.
Tackle theorems and proofs with determination,
For in the world of Calculus lies endless fascination.

Let your curiosity soar like a bird in the sky,
As you uncover the truths that Calculus implies.

The path may be winding, the concepts complex,
But with each step forward, you'll conquer the vex.

Through tangents and gradients, you'll gain insight,
Into the patterns and rhythms that make the world bright.
In Calculus, lies the power to understand,
The universe's language, written in your hand.

So, fear not the challenges that lie ahead,
For with diligence and focus, you'll forge ahead.
With each equation conquered, each problem solved,
The rewards of success shall surely evolve.

You have the strength within, the spirit to persist,
To conquer Calculus and top the highest list.
Embrace the subject, explore its grand design,
And witness the boundless flight of knowledge, divine.

THREE

FEAR NOT

In the realm of numbers, where mystery lies,
There's a subject that challenges, reaching the skies.
Calculus, they call it, a language of change,
With curves and equations, it may seem strange.

Fear not, dear student, for within its domain,
Lies a world of beauty, waiting to be claimed.
From limits to derivatives, it may seem tough,
But the rewards it offers are more than enough.

Through integration, secrets are revealed,
Of motion, of growth, of patterns concealed.
It unlocks the mysteries of the universe's core,
And empowers you to explore even more.

With pencil in hand, and a mind open wide,
Dive into the depths, let curiosity be your guide.

Embrace the challenges, for they make you grow,
And with perseverance, success you'll surely know.
　　Though the road may be winding, and the journey long,
Remember, dear student, you are strong.
With each step you take, you'll conquer the heights,
And emerge victorious, bathed in the brightest lights.
　　So fear not, my friend, for you hold the key,
To unlock the wonders of Calculus, you see.
Embrace the subject, with all your might,
And soar to new heights, like an eagle in flight.

FOUR

YOU'LL CONQUER

In the realm of numbers, where mysteries reside,
Lies a subject called Calculus, a thrilling ride.
Fear not, dear student, for I'll be your guide,
Through this mathematical journey, side by side.

Like a mountain to conquer, it may seem vast,
But with perseverance, you'll reach it at last.
Calculus unlocks secrets, like a hidden key,
Revealing the language of nature's decree.

From derivatives to integrals, it unveils the unknown,
Equations dance with joy, as truths are shown.
Through limits and functions, the world comes alive,
In the intricate web that Calculus does derive.

Embrace the challenge, let curiosity ignite,
As you delve into the depths, with all your might.

For within the equations, beauty you will find,
A symphony of patterns, a masterpiece of the mind.
   Let Calculus be your compass, guiding you through,
A voyage of discovery, where dreams come true.
So don't be discouraged, let passion burn bright,
And with dedication, you'll conquer the height.
   For in the realm of numbers, where mysteries reside,
Lies a subject called Calculus, your soul's true guide.
Unlock the wonders, let knowledge take flight,
And soar through the realms, with Calculus as your light.

FIVE

PASSIONATE PURSUIT

In the realm of numbers, where logic abides,
Lies a subject that challenges, where brilliance resides.
Calculus, the language of change and motion,
Unveils the secrets of nature with deep devotion.
    Fear not the symbols, the equations daunting,
For within their complexity, lies a world enchanting.
Calculus, the gateway to understanding the unseen,
Unravels the mysteries that lie in between.
    Embrace the challenge, embrace the strife,
For through perseverance, you'll unlock life.
Calculus, the key to unraveling the unknown,
Guides you through the darkness, to the light shone.
    With limits and derivatives, you'll learn to explore,
The depths of functions, their curves to adore.

Calculus, the art of dissecting the infinite,
Reveals patterns and behaviors, with every minute.
   So, fear not the integration, the sums and the sums,
For within their complexities, success surely comes.
Calculus, the language of progress and growth,
Leads you to triumph, as you embrace both.
   Through differentiation and integration's embrace,
You'll conquer the challenges, each step you'll trace.
Calculus, the path to knowledge and power,
Unleashes your potential, in every hour.
   So, dear student, let not fear hold you back,
Embrace the beauty of Calculus, stay on track.
For within its realm, wonders await,
Passionate pursuit, your destiny, you'll create.

SIX

DEAR STUDENT

In the realm of numbers, where logic prevails,
Lies a subject that often makes hearts quail.
Calculus, the language of curves and change,
A journey that may seem daunting and strange.

But fear not, dear student, for within its folds,
Lie treasures untold, secrets yet to be told.
Like a compass guiding you through the unknown,
Calculus shall be your beacon, your very own.

With limits and derivatives, it unveils,
The mysteries of motion, how nature prevails.
From the arc of a rocket to the shape of a wave,
Calculus unravels the patterns that pave.

It's the language of growth, of rates and trends,
A powerful tool, on which science depends.
Through integrals, it finds areas untamed,
Drawing connections where none could be named.

So let your curiosity ignite, like a flame,
Embrace the challenges, let Calculus claim.
For within its depths, you'll find treasures anew,
A world of knowledge, waiting for you.

Persevere through the theorems, the equations so vast,
And you'll conquer this subject, a triumph unsurpassed.
Unlock the secrets of the universe's design,
With Calculus as your guide, your brilliance will shine.

So fear not the numbers, the symbols, the signs,
For Calculus is a journey, where greatness aligns.
With dedication and focus, you'll surely surpass,
And witness the beauty of Calculus, en masse.

SEVEN

POWER TO CREATE

In the realm of numbers, Calculus resides,
A subject that beckons, where wonders hide.
With derivatives and integrals, it unfolds,
A language of patterns, a story untold.

 Fear not, O student, for within lies the key,
To unlock the secrets, to set your mind free.
Calculus, a gateway to knowledge profound,
A realm of understanding where beauty is found.

 From limits to functions, it takes you afar,
To explore the universe, a celestial memoir.
The curves and slopes, they dance in delight,
Revealing the secrets of day and of night.

 Embrace the challenge, let courage arise,
For every hurdle, a reward in disguise.

With each equation, a puzzle to solve,
A journey of growth, where strength will evolve.
   In Calculus lies the power to create,
To model the world and its intricate state.
From physics to finance, it finds its place,
A tool of discovery, a path to embrace.
   So, dear student, let not your spirit wane,
In the face of difficulties, do not refrain.
For in the realm of Calculus, brilliance resides,
Unlock its wonders, let your dreams take flight.

EIGHT

MASTERING CALCULUS

In the realm of numbers, where logic intertwines,
There lies the realm of Calculus, where brilliance shines.
Through limits and derivatives, it takes flight,
A subject that challenges with all its might.

Fear not, dear student, for you hold the key,
To unravel the secrets of this mystery.
Calculus, like a puzzle waiting to be solved,
Reveals the wonders waiting to be evolved.

With integrals and functions, you'll dance in delight,
As the curves and slopes come into clear sight.
Embrace the challenges, let your spirit be bold,
For within the depths of Calculus, treasures unfold.

It may seem daunting, with equations so complex,

But with patience and practice, you'll conquer each test.
Every step forward, no matter how small,
Brings you closer to understanding it all.
 So let your passion ignite, let your mind soar,
Unlock the realms of Calculus like never before.
For in this subject lies a power untold,
A gateway to knowledge, a future to behold.
 With diligence and perseverance, you'll surely succeed,
In mastering Calculus, fulfilling your academic creed.
So fear not, dear student, for this journey's worth the strife,
For in the world of Calculus, lies a beautiful life.

NINE

SUCCESS IS YOURS

In the realm of numbers and lines,
Where patterns hide and knowledge shines,
Lies a subject that beckons, calls,
The wondrous world of Calculus.

Fear not the symbols, strange and bold,
For they hold secrets yet untold.
Through limits, derivatives, and more,
A universe of wonder lies in store.

Like a puzzle waiting to be solved,
Calculus reveals what's involved.
It unlocks the mysteries of change,
And shows the way to rearrange.

Embrace the challenge, don't shy away,
For in the struggle, you'll find your way.

With curiosity as your guide,
The beauty of Calculus will never hide.
    Persevere, through highs and lows,
With dedication, the knowledge grows.
Practice, practice, day by day,
And mastery will come your way.
    Unlock the secrets, set your mind free,
For Calculus is a door, you see.
To understanding the world around,
And the wonders that abound.
    So fear not the symbols, embrace the unknown,
In Calculus, your skills will be honed.
With patience, practice, and a fearless spirit,
Success is yours, don't fear to hear it.
    Pass the subject, conquer the test,
And know that you've done your very best.
For in the journey, you'll come to find,
That Calculus is truly one of a kind.

TEN

PATH TO GREATNESS

In the realm of Calculus, where numbers dance and sway,
Lies a world of wonder, where insights hold their sway.
Fear not the symbols, the equations that confound,
For within those mysteries, true brilliance can be found.

Calculus, the language of the universe's design,
Unveils the secrets of motion, patterns so divine.
From the curves of nature to the stars in the night,
It unravels mysteries, casting darkness into light.

With limits and derivatives, we explore the unknown,
Unleashing our potential, a journey all our own.
Through integrals and functions, we grasp the grand design,

Unlocking the power of knowledge, a treasure so sublime.

Embrace the complexity, let curiosity ignite,
For in the depths of Calculus, new worlds take flight.
Persevere through challenges, for they hold the key,
To mastering this subject, and setting your mind free.

So, dear student, heed my words, let them inspire,
Let Calculus be your guide, fuel your burning desire.
For within its realm, lies a path to greatness untold,
Passion, dedication, and a future that unfolds.

ELEVEN

YOU'LL SURELY PASS

In the realm of Calculus, where numbers dance,
A student seeks to take a chance.
Fear not the symbols, the theorems, and the signs,
For within their depths, true treasure shines.

Embrace the challenge, let curiosity ignite,
For in the world of Calculus, wonders take flight.
Like a mathematician, a pioneer you'll be,
Unraveling mysteries, setting your spirit free.

With every integral, a puzzle you'll unfold,
Finding solutions, more valuable than gold.
Differentiation, a tool to understand,
The changing world, both vast and grand.

Persevere through limits, as they test your might,
For with dedication, you'll conquer any height.
The beauty of Calculus lies in its grace,
A language that unveils the universe's embrace.

Practice, practice, day and night,
Let your passion burn, shining bright.
With each problem solved, knowledge will grow,
And success in Calculus, you shall bestow.

So fear not, brave student, for you have the power,
To conquer Calculus, hour by hour.
With diligence and faith, you'll surely pass,
And in the realm of numbers, find your class.

TWELVE

SETTING YOU FREE

In the realm of numbers and curves,
Where logic and mystery converge,
Lies the wondrous realm of Calculus,
An art that may seem treacherous.
    Fear not the limits and derivatives,
For within them, knowledge lives.
With every integral and tangent line,
New insights and truths you'll find.
    Calculus, a language of change,
Unraveling patterns, rearranging the range.
It holds the power to solve life's puzzles,
To understand motion and its subtle nuances.
    Embrace the challenge, take a leap,
And in its depths, your spirit keep.

For though it may seem daunting now,
With persistence, you'll learn to plow.
    Through functions, equations, and rates of change,
A world of possibilities will arrange.
Discover the beauty that lies beneath,
The elegance of Calculus, a bittersweet relief.
    So, student, do not shy away,
From Calculus, come what may.
With practice, dedication, and a fearless mind,
Success in this subject, you will surely find.
    For in the journey of learning, you'll see,
The power of Calculus, setting you free.
Unlock its secrets, embrace the strife,
And conquer the subject that brings wisdom to life.

THIRTEEN

KNOWLEDGE YOU INVEST

In the realm of numbers, where logic resides,
Lies a subject that challenges, where fear collides.
Calculus, they call it, a mountain to ascend,
Where equations dance, and limits transcend.

At first glance, it may seem daunting and tough,
With derivatives and integrals, more than enough.
But fear not, dear student, for within this maze,
Lies the key to unlock countless unseen ways.

Calculus is a gateway, a portal to explore,
A language of nature, where mysteries galore.
From the motion of planets to the growth of a tree,
It unravels the secrets, for all to see.

Every curve and tangent, every area and rate,
Calculus unveils their stories, it's never too late.

So, embrace the challenge, let your spirit soar,
For in conquering Calculus, you'll crave for more.
 With each problem solved, a victory is won,
A step closer to mastery, a battle hard-fought and won.
So, persevere, my student, let your passion ignite,
For in the realm of Calculus, your future shines bright.
 Remember, it's not just about passing the test,
It's about the journey, the knowledge you invest.
So, dive deep into the depths, let your passion ignite,
And Calculus will reward you, with wisdom and light.

FOURTEEN

CONQUER CALCULUS

In the realm of numbers, where mysteries lie,
There's a subject that challenges the mind's eye.
Calculus, they call it, a world of its own,
Where limits and derivatives have seeds sown.

    Fear not, young student, for I bring a tale,
Of perseverance and knowledge that shall never fail.
Embrace this calculus, like a mountain to climb,
For within its depths, your true potential will shine.

    With Euler's method and integrals to explore,
Let passion and dedication be your core.
For every theorem and concept you grasp,
A new horizon of understanding you'll clasp.

    Like a river that carves through the rough terrain,
Calculus reveals truths that will never wane.

It's not about shortcuts or mere tricks,
But the pursuit of wisdom that truly sticks.
   Let the challenges fuel your desire,
To conquer Calculus with a burning fire.
For through perseverance and relentless stride,
Success will be yours, with Calculus as your guide.
   So, fear not the symbols, the equations, the tests,
For within your heart, lies the power to invest.
In the realm of Calculus, a journey awaits,
Where growth and achievement will be your fate.

FIFTEEN

GROWTH AND ACHIEVEMENT

In the realm of Calculus, a mighty force does dwell,
A world of numbers, equations, where wisdom does excel.
Fear not this daunting subject, embrace it with delight,
For within its depths lies beauty, a beacon shining bright.
    Calculus, the language of change, a powerful tool indeed,
Unlocking the secrets of motion, patterns, and speed.
With limits and derivatives, it charts the course of the unknown,
Revealing the intricate dance of curves, a symphony of its own.

Though challenges may arise, and frustrations may be found,
Know that with practice and dedication, success will abound.
Let not the fear of failure hinder your journey's start,
For it's through perseverance and determination that you'll leave your mark.

Remember, it's not just about passing a test or earning a grade,
But about the growth and knowledge gained along the way.
In this pursuit of understanding, you'll find a world of power,
A universe of logic and reasoning, where dreams can truly flower.

So dive deep into the realm of Calculus, let your passion ignite,
Embrace the challenges that come, and face them with all your might.
For shortcuts and tricks can only take you so far,
But the pursuit of wisdom and understanding will make you a shining star.

Believe in yourself, dear student, and let your heart invest,
In this wondrous realm of Calculus, where growth and achievement rest.
With perseverance and dedication, success will be your claim,

For within your very being lies the power to conquer this game.

SIXTEEN

WONDROUS FLIGHT

In the realm of numbers, where equations thrive,
Lies a subject that tests our will to survive.
Calculus it's called, a mountain to climb,
But fear not, dear student, for I'll share a rhyme.

    Let's embrace the challenges, the trials we face,
For within them lies the beauty and grace.
The limits we find, the derivatives we seek,
They hold the secrets we long to peek.

    With perseverance and dedication, we'll soar,
Unlocking the mysteries, forever wanting more.
Integration, differentiation, and all in between,
The power of Calculus, let it be seen.

    Fear not the symbols, the theorems, or signs,
For deep within, the answers you'll find.

Believe in yourself, in your boundless might,
And Calculus will be a wondrous flight.
    Embrace the concepts, let curiosity guide,
As you delve deeper, your fears will subside.
For in this journey, knowledge shall bloom,
And success in Calculus shall be your own.
    So, dear student, take heart and take heed,
Calculus is not just about the test you'll succeed.
It's about growth, learning, and the power you'll find,
To conquer the subject and expand your mind.

SEVENTEEN

CURVES AND SLOPES

In the realm of numbers, a kingdom vast,
Lies the subject of Calculus, a journey to embark,
Fear not, dear student, for I shall unveil,
A world of beauty, where knowledge imparts.

Amid the curves and slopes, equations dance,
Like melodies composed by a masterful hand,
Calculus, the language of change and chance,
Unfolding the secrets of the universe's grand.

From derivatives to integrals, we delve,
Into the depths of limits and infinite sum,
Each concept a puzzle, waiting to be solved,
By your determination, you shall overcome.

Embrace the challenges, let curiosity guide,
For in the struggles, wisdom does reside,

Push your boundaries, expand your mind,
Through Calculus, new perspectives you'll find.
   It's not just about passing a test or a grade,
But about the growth that learning has made,
For in the pursuit of knowledge, you'll see,
The power within you, waiting to be set free.
   So, dear student, take heart and believe,
In the wonders of Calculus you shall receive,
Embrace the concepts, let your passion ignite,
For within yourself, the answers lie in sight.

EIGHTEEN

PURE DELIGHT

In the realm of numbers, where mystery resides,
Lies the subject of Calculus, where knowledge abides.
Fear not, dear student, for I bring you a tale,
Of conquering challenges, where victory prevails.

Calculus, a beast with teeth so sharp,
Yet its beauty lies in the depths of its art.
Like a puzzle unsolved, waiting to be unraveled,
It beckons you, student, with problems to grapple.

Embrace the challenge, let your passion ignite,
For within the numbers, a world takes flight.
Equations and functions, they dance and they sway,
Revealing secrets that will light up your way.

Dive deep into limits, explore the unknown,
Discover the power that Calculus has shown.

From derivatives to integrals, every concept unfolds,
Unveiling the wisdom that the subject beholds.

It's not just about passing a test, my friend,
But about growth, knowledge, and the power to transcend.
With dedication and perseverance, you shall prevail,
And the gates of understanding, you'll surely unveil.

So fear not the numbers, let your spirit soar,
For success in Calculus will open new doors.
Believe in yourself, for you hold the key,
To unlocking the wonders of this subject, you see.

Let your passion guide you, let your dreams take flight,
For in the realm of Calculus, you'll find pure delight.
So embrace the concepts, let your knowledge expand,
And witness the world through Calculus's grand hand.

NINETEEN

WITNESS THE MIRACLES

In the realm of numbers, where Calculus resides,
A student embarks on a wondrous ride.
With equations and limits, they'll soon come to see,
The boundless potential that lies within thee.

Fear not the challenges that Calculus brings,
For with dedication, you'll soar on its wings.
Embrace the derivatives, the integrals, too,
And watch as your knowledge begins to accrue.

Like a puzzle, Calculus unveils its art,
As you delve into functions, each piece a fresh start.
Let curiosity guide you, like a compass in hand,
And discover the secrets, like grains in the sand.

Believe in your abilities, for they are vast,
With perseverance, you'll conquer the past.

Through each trial and error, you'll learn and grow,
Unveiling the wisdom only Calculus can bestow.

For Calculus is not just about passing a test,
It's about understanding, the journey, the quest.
It's the power to transcend, to reach for the stars,
And unlock the wonders that lie beyond the bars.

So fear not, dear student, for you hold the key,
To unlock the beauty that Calculus can be.
Embrace the challenges, let knowledge unfold,
And watch as your brilliance begins to behold.

For in the depths of Calculus, you'll find,
A world of knowledge that expands your mind.
So study, persevere, and never relent,
And witness the miracles Calculus presents.

TWENTY

SELF-TRANSCEND

In a realm of numbers, Calculus stands,
Where equations dance and minds expand.
Fear not the challenges that lie ahead,
For in perseverance, greatness is bred.
    Embrace the limits, both finite and vast,
Let curiosity guide you steadfast.
With dedication, your path will unfold,
Unveiling the wonders Calculus holds.
    From limits to derivatives, explore the unknown,
Unlock the secrets that lie yet unknown.
Integration's embrace, a symphony grand,
Revealing truths as you take a firm stand.
    Calculus, a language of nature's design,
Revealing patterns, both subtle and fine.

Through tangents and curves, a story unfolds,
A tale of growth, where wisdom takes hold.
  Believe in your abilities, strong and true,
For Calculus is the key to breakthrough.
With every integral and derivative found,
You'll soar to heights, with knowledge profound.
  So, student, heed this poetic plea,
Let Calculus be your gateway to glee.
Pass not just a test, but your own self-transcend,
And witness the miracles Calculus can send.

TWENTY-ONE

REACH FOR THE SKIES

In the realm of numbers, a language profound,
Lies the secret to nature, where wonders abound.
Calculus, the gateway to infinite might,
Unveils the mysteries hidden from sight.

Fear not the symbols, nor the theorems grand,
For within lies a power, ready to expand.
Let your passion ignite, let it soar and ignite,
And witness the miracles, as day turns to night.

Embrace the challenge, let curiosity bloom,
For in every equation, there's wisdom to consume.
With every derivative and integral you take,
A new world unfolds, like a shimmering lake.

Calculus, a journey, not just a test to pass,
A pathway to growth, where your mind shall amass.

Unlock the secrets, let your mind roam free,
And witness the beauty that others cannot see.

Through limits and functions, you'll transcend the norm,
Unlocking new doors, your mind will transform.
Let Calculus guide you, as you reach for the skies,
And witness the power that within you lies.

With dedication and perseverance, you shall find,
The strength to conquer, to leave no task behind.
So fear not, dear student, for success is in sight,
Embrace Calculus' wonders, and let your knowledge ignite.

TWENTY-TWO

UNDERSTANDING CALCULUS

In the realm of numbers, where mysteries reside,
Lies a subject called Calculus, a wondrous stride.
It beckons you, dear student, to unlock its gate,
To embrace its challenges and conquer your fate.

Calculus, a language of motion and change,
A symphony of derivatives, a dance so strange.
It's a path to understanding this world's design,
A tool to unravel the secrets of space and time.

Fear not the limits and integrals that lie ahead,
For within their complexities, wisdom is spread.
They teach you to analyze, to reason, to prove,
To see the world through a lens of infinite groove.

As you delve into functions, curves, and slopes,
You'll find beauty in patterns, where knowledge

elopes.

The power of Calculus lies in its embrace,
To explore the unknown and expand your own space.

So, hold on to your courage, let doubts fall away,
With each equation you conquer, you'll grow day by day.
Embrace the challenges, for they make you strong,
And remember, dear student, you've had it in you all along.

Passing the test is just the first step in the game,
But understanding Calculus will set your soul aflame.
Embrace its power, its beauty, its might,
And soar to new heights, in Calculus' insightful light.

TWENTY-THREE

EXPAND YOUR HORIZONS

In the realm of numbers, let your spirit take flight,
Embrace the challenge and conquer with might,
For Calculus holds the key to the unknown,
A pathway to growth, like seeds that are sown.

Beyond the equations lies a world untold,
Where limits are pushed and wonders unfold,
Let curiosity guide you, let it be your muse,
As you navigate the terrain, don't be afraid to choose.

Calculus is not just numbers on a page,
It's a language of nature's design, an intellectual stage,
It reveals the secrets of motion and change,
Unveiling the mysteries that rearrange.

Through integrals and derivatives, you will see,
The power to solve problems, set your mind free,

Unlock the doors to new realms of thought,
Expand your horizons, let your dreams be sought.
    Believe in yourself, in your ability to learn,
For knowledge is the fire that will forever burn,
With Calculus as your guide, you'll reach for the stars,
And witness the miracles that lie beyond the bars.
    So embrace the power, the beauty, the grace,
Let your knowledge ignite and transform your space,
Passing Calculus is not just a test,
It's a journey of understanding, a chance to be your best.

TWENTY-FOUR

BOUNDLESS SEA

    In the realm of numbers, a journey waits,
Where Calculus unveils its sacred gates.
Oh student, listen close and hear my plea,
Embrace this art, unfold your destiny.
    Through limits, derivatives, and integrals grand,
A world of patterns, surreal and unplanned.
Discover the power to shape the unknown,
Where equations dance, in harmony they're sown.
    Fear not the challenges that lie ahead,
For in each hurdle, growth is gently bred.
Like a mountain climber scaling new heights,
Calculus will elevate your insights.
    From curves that soar and tangents that kiss,
To volumes and areas you won't want to miss.
Unlock the secrets of this mystic art,
And witness the wonders that lie at its heart.

The beauty of Calculus, like a cosmic dance,
Unfolding the universe with every chance.
In its embrace, you'll find wisdom's key,
A pathway to self-transcendence, you'll see.

So, student, take heart, and cast away doubt,
For understanding awaits, there's no need to pout.
Passing a test is just the beginning, you see,
The real triumph lies in unlocking the key.

Embrace the challenge, reach for new heights,
Let Calculus ignite your inner lights.
For in this subject lies a boundless sea,
Where growth, knowledge, and power shall forever be.

TWENTY-FIVE

BRILLIANCE EXPRESSED

In the realm of numbers and equations,
Lies a subject that sparks fascination,
Calculus, the art of change and motion,
A gateway to knowledge and deep devotion.

Oh, student, hear my fervent plea,
Embrace this challenge with all your glee,
For within the bounds of Calculus' might,
Lies the power to illuminate your sight.

Discover the secrets of curves and lines,
As you solve problems, your spirit shines,
From limits to derivatives, unleash your mind,
And in the realm of Calculus, wisdom you'll find.

Passing a test is just the start,
For Calculus offers a world apart,

A journey of growth, a path of grace,
Where you'll transcend and find your place.
    Believe in yourself, oh student dear,
Unlock your potential, banish the fear,
Calculus is not just numbers on a page,
It's a symphony of knowledge, a wisdom sage.
    So study hard, and don't despair,
For in your heart, the answers are there,
With determination, you'll conquer the test,
And soar to new heights, your brilliance expressed.
    Embrace Calculus, let it be your guide,
A subject of beauty, where dreams reside,
Pass the test, and then you'll see,
The wonders of Calculus, a key to set you free.

TWENTY-SIX

POSSIBILITIES AND PARADOX

In the realm of numbers and shapes,
A subject of beauty, Calculus takes shape.
Fear not the challenges it may impose,
For within lies the path to growth it bestows.
    With limits and derivatives, we explore,
The power to understand and so much more.
Equations unravel their secrets untold,
As we embrace the wonders Calculus holds.
    Through integrals, we measure the vast,
The curves of life, forever they last.
From moments to areas, it reveals,
The hidden truths that the world conceals.
    With each calculation, a door unlocks,
A gateway to possibilities and paradox.

Embrace the complexity, let it unfold,
For in its depths, your dreams will be told.
 Passing the test, just the beginning it seems,
For Calculus offers so much more it gleams.
A journey of self-transcendence it unveils,
A path to reach your highest potential, it entails.
 Believe in yourself, let doubt be erased,
With perseverance and dedication, you'll embrace,
The profound beauty that Calculus imparts,
And unlock the secrets of infinite arts.
 So, dear student, let Calculus inspire,
Ignite your passion, set your dreams on fire.
For in this subject lies a world unseen,
Where the magic of numbers reigns supreme.

TWENTY-SEVEN

YOUR MIND UNFOLD

In the realm of numbers, where mysteries reside,
Lies a subject that many deem hard to abide.
Calculus, the enchantress of mathematical art,
A gateway to knowledge, a journey to embark.

Fear not the symbols, the equations that gleam,
For within them lies a cosmic dance, it may seem.
Limits and derivatives, they hold the key,
To unlocking the secrets of the world's decree.

Oh, student of wonder, with dreams to pursue,
Calculus beckons, offering something new.
Embrace the challenge, let your spirit soar,
For within its depths, you'll discover much more.

From slopes and tangents to curves that bend,
Calculus unravels the fabric, my friend.

It weaves together the fabric of space and time,
Unveiling the mysteries, so sublime.

   So study with diligence, persevere with might,
For in conquering Calculus, your future shines bright.
Believe in yourself, in your potential, you'll find,
That passing the test is just the first step in kind.

   For Calculus offers a journey, a quest,
To transcend your limits and be your very best.
Unlock its secrets, let your mind unfold,
And watch as new horizons, your future behold.

   In the realm of numbers, where dreams take flight,
Calculus awaits, with wonders so bright.
So fear not, dear student, for you have what it takes,
To conquer the subject, and the world, it awakes.

TWENTY-EIGHT

PERSEVERANCE AND A CURIOUS MIND

In the realm of numbers, you'll find a treasure,
A subject that unlocks paths beyond measure.
Calculus, the language of change and motion,
Holds the power to ignite your true devotion.

Embrace the challenge, let your mind take flight,
For in the depths of Calculus, you'll find pure delight.
Limits and derivatives, they may seem tough,
But through dedication, you'll rise high enough.

Like a painter with a brush, you'll create,
Graphs and equations, a masterpiece awaits.
From curves that dance with elegance and grace,
To slopes that guide us through time and space.

Integrals, the puzzle pieces we seek,
Solving them, you'll reach the mountain's peak.

Area, volume, and rates of change,
Calculus opens doors, it's truly strange.
    So fear not the symbols, the theorems, and rules,
In this realm of numbers, you'll never be fools.
With perseverance and a curious mind,
You'll conquer Calculus, leaving doubts behind.
    Unlock the door to boundless possibilities,
Through Calculus, you'll discover your capabilities.
So dive deep into the world of math,
And let your passion guide you on the path.
    Believe in yourself, for you have the power,
To conquer Calculus, hour by hour.
With every step, you'll grow and learn,
And triumph over challenges, it's your turn.

TWENTY-NINE

MUCH TO GAIN

In the realm of numbers, where mysteries reside,
Lies a subject profound, where wonders abide.
Calculus, a language of the infinite and grand,
Unveils the secrets of nature, held within its hand.

Embrace the power, the beauty it holds,
For in its depths, a story unfolds.
From limits and derivatives, a symphony of change,
To integration, where patterns rearrange.

Like a master architect, you'll learn to build,
Equations and functions, with skill and strong will.
The curves and slopes, they dance with grace,
Revealing the marvels of time and space.

Fear not the challenges that lie ahead,
For with every hurdle, you'll grow and spread.
Through differentiation, you'll find your stride,

And conquer the subject, with knowledge as your guide.

Trust in yourself, in your brilliance and might,
With diligence and perseverance, you'll shine bright.
Unlock the boundless sea of knowledge and growth,
And witness the wonders that Calculus bestows.

So, dear student, let not despair take hold,
Embrace the triumphs, both new and bold.
For in the world of Calculus, there's much to gain,
A world of infinite possibilities, ready to entertain.

THIRTY

SOLVE THE PUZZLES

In realms where numbers intertwine,
Where curves and lines forever shine,
There lies a realm of wondrous grace,
Where Calculus unveils its face.

Within its depths, a power lies,
To solve the puzzles of the skies,
To understand the grand design,
Of nature's laws, both subtle and fine.

Fear not the symbols that may confound,
For in their midst, brilliance is found.
Derivatives, integrals, they may seem,
Like a daunting, perplexing dream.

But fear not, dear student, for I say,
With perseverance, find your way.

Unlock the secrets, one by one,
And see the beauty that's just begun.
    Calculus, a language of the wise,
A gateway to the boundless skies.
Embrace its challenge, seize the chance,
To transcend limits, to enhance.
    Believe in yourself, let doubts be gone,
For in your heart, the strength is strong.
Through trials and errors, you will find,
A world of knowledge, vast and kind.
    So let your passion guide you through,
The realms of Calculus, bold and true.
For within its depths, you'll come to see,
The endless possibilities.

THIRTY-ONE

EMBRACE THE CHALLENGE

In the realm of numbers, where equations dance,
Lies the essence of knowledge, a captivating trance.
Calculus, a gateway to realms unknown,
A subject that carves a mind into a throne.
From limits to derivatives, it may seem tough,
But fear not, dear student, for you're made of tough stuff.
Embrace the challenge, let your curiosity bloom,
For in the depths of Calculus, wonders loom.
With each integral solved and each problem cracked,
A realm of understanding will be unpacked.
Like a sculptor, chiseling away the excess,
Calculus shapes your mind, leaving it blessed.

It's not just about grades, it's about the quest,
To unravel the mysteries, to be your very best.
The beauty lies not in the final conclusion,
But in the process, the growth, the self-transfusion.

So, fear not the symbols, the theorems, the signs,
For within them, a world of discovery shines.
Let Calculus be your guide, your compass, your light,
And watch as it unveils infinite possibilities in sight.

Believe in yourself, for you hold the key,
To conquer the challenges, to set your mind free.
For Calculus is not just a subject to pass,
It's a journey of growth, a chance to amass.

So, dear student, let your passion ignite,
Embrace Calculus, reach for the infinite height.
With perseverance and dedication, you shall surpass,
And in the realm of Calculus, forever you'll class.

THIRTY-TWO

BELIEVE IN YOURSELF

In the realm of numbers and curves,
Where the mind dances and swerves,
Lies the gateway to endless heights,
A realm where knowledge takes flight.
    Oh student, hear my humble plea,
Embrace the wonders of Calculus, you'll see,
It's not just equations and rules,
But a language that unlocks the jewels.
    From derivatives to integrals,
Calculus weaves its magical spells,
Unraveling the mysteries of change,
In every function, it rearranges.
    With limits and continuity,
Calculus unveils its serenity,

It challenges your very core,
But it's a journey worth fighting for.

    Through the valleys of curves so steep,
You'll learn to conquer and leap,
For every challenge that comes your way,
Builds strength to face another day.

    So don't despair, my student dear,
Calculus may at times bring fear,
But with perseverance and steadfast will,
You'll conquer each summit, standing still.

    Believe in yourself, the path is clear,
Calculus is the bridge, my dear,
To a world of knowledge and endless growth,
Where your potential truly shows.

    So let your mind soar high and free,
Embrace the beauty of Calculus, you'll see,
Passing the test is just the start,
Unleashing your brilliance, a work of art.

THIRTY-THREE

CHANCE TO OVERCOME

In the realm of numbers, where Calculus dwells,
A student embarks on a journey, where wonders are unveiled.
Fear not the complexities that lie ahead,
For with perseverance, you shall conquer instead.
　　Calculus, a language of curves and equations,
A treasure trove of mathematical revelations.
Let curiosity guide you through its intricate maze,
And watch as your understanding slowly starts to blaze.
　　Embrace the challenge, for it holds the key,
To unlock the secrets of the universe, you see.
With every derivative and integral you take,
A new perspective on the world will awake.
　　Fear not the limits and the infinite sums,

For in every problem, lies a chance to overcome.
With each step forward, you'll grow and evolve,
As Calculus helps your mind to revolve.

So, sharpen your pencils and open your mind,
Let the beauty of numbers and logic unwind.
Remember, it's not just a subject to pass,
But a journey of growth and self-discovery, alas.

Believe in yourself, for you hold the power,
To conquer Calculus, hour after hour.
Embrace the challenge, let your passion ignite,
And watch as your dreams take flight.

THIRTY-FOUR

REWARDS OF MASTERING CALCULUS

In the realm where numbers dance and soar,
Lies a subject called Calculus, knocking at your door.
Fear not, dear student, for within its depths,
Lies a world of wonders, where knowledge intercepts.

Let your heart be bold, your mind ablaze,
For Calculus is but a puzzle to embrace.
Its language may seem foreign, its concepts complex,
But with perseverance, you'll conquer and progress.

Through limits and derivatives, you'll find your way,
Unraveling the mysteries, day by day.
For every integral you solve with might,
A universe of possibilities takes flight.

Embrace the challenges, let passion guide,
For in this journey, your potential will be amplified.
Let the curves and slopes ignite your mind,
As you explore the boundless depths you'll find.
Believe in yourself, for you hold the key,
To unlock the secrets of Calculus, and set yourself free.
With each hurdle you overcome, your strength will grow,
Leading you to heights you never thought you'd know.
So fear not, dear student, for you have the power,
To conquer Calculus and bloom like a flower.
Embrace the journey, with a curious mind,
For the rewards of mastering Calculus are one of a kind.

ABOUT THE AUTHOR

Walter the Educator is one of the pseudonyms for Walter Anderson. Formally educated in Chemistry, Business, and Education, he is an educator, an author, a diverse entrepreneur, and he is the son of a disabled war veteran. "Walter the Educator" shares his time between educating and creating. He holds interests and owns several creative projects that entertain, enlighten, enhance, and educate, hoping to inspire and motivate you.

Follow, find new works, and stay up to date
with Walter the Educator™
at WaltertheEducator.com

www.ingramcontent.com/pod-product-compliance
Lightning Source LLC
LaVergne TN
LVHW052001060526
838201LV00059B/3772